FOSSIL RIDGE PUBLIC LIBRARY DISTRICT
W9-BRA-603

Energy in Action™

ELECTRICITY

The Rosen Publishing Group's
PowerKids Press™
New York

Ian F. Mahaney

Published in 2007 by the Rosen Publishing Group, Inc.
29 East 21st Street, New York, NY 10010

First Edition

Editor: Joanne Randolph
Book Design: Julio Gil

Photo Credits: Cover, title page © Richard Cummins/Corbis; cover and interior pages (top right) © www.istockphoto.com/Michelle Van Meter; p. 5 © Scott Stulberg/Corbis; p. 10 © Little Blue Wolf Productions/Corbis; pp. 11, 18 © Michael S. Yamashita/Corbis; p. 12 © Lawrence Manning/Corbis; p. 13 © Lester Lefkowitz/Corbis; p. 15 © Roy McMahon/Corbis; p. 16 © Werner H. Müller/Corbis; p. 17 © www.istockphoto.com/Igor Karon; p. 19 © George Steinmetz/Corbis; pp. 20, 21 Adriana Skura; p. 22 Cindy Reiman.

Library of Congress Cataloging-in-Publication Data

Mahaney, Ian F.
Electricity / Ian F. Mahaney.— 1st ed.
p. cm. — (Energy in action)
Includes index.
ISBN (10) 1-4042-3478-0 (13) 978-1-4042-3478-9 (lib. bdg.) —
ISBN (10) 1-4042-2187-5 (13) 978-1-4042-2187-1 (pbk.)
1. Electricity—Juvenile literature. 2. Electrostatics—Juvenile literature. 3. Electric power—Juvenile literature. I. Title. II. Energy in action (PowerKids Press)
QC527.2.M337 2007
537—dc22

2005035719

Manufactured in the United States of America

CONTENTS

Energy

Having energy means you can do active things. Playing at recess, going to the beach, shoveling the snow, or mowing the lawn all take energy. If you have energy, you have the ability to accomplish something.

In science energy is the ability to do work. There are many kinds of energy. These include **chemical** energy, **mechanical** energy, **thermal** energy, and electrical energy. Energy is being used or created everywhere in nature. Heat given off from the Sun warms Earth. The wind cools us down. Lightning is a form of energy, too. It is natural **electricity**. People make electricity, too. Let's take a closer look at electrical energy.

Lightning is a very powerful form of natural electricity. The energy in a bolt of lightning heats the air through which it passes to more than 50,000° F (27,760° C). More than 20 million bolts of lightning hit the ground in the United States each year.

What Is an Atom?

To understand electricity we need to understand the atom. Atoms are the tiny parts that make up everything on Earth. Your pencil is made of atoms and so are your fingers.

Every atom has three parts, called **neutrons**, **electrons**, and **protons**. The protons and neutrons make up the central part of the atom. This part is called the nucleus. The electrons spin around the nucleus. The three parts of an atom have what is called a **charge**. The different charges of the different parts of the atom cause them to attract or repel one another. "Attract" means "pull toward," and "repel" means "push away."

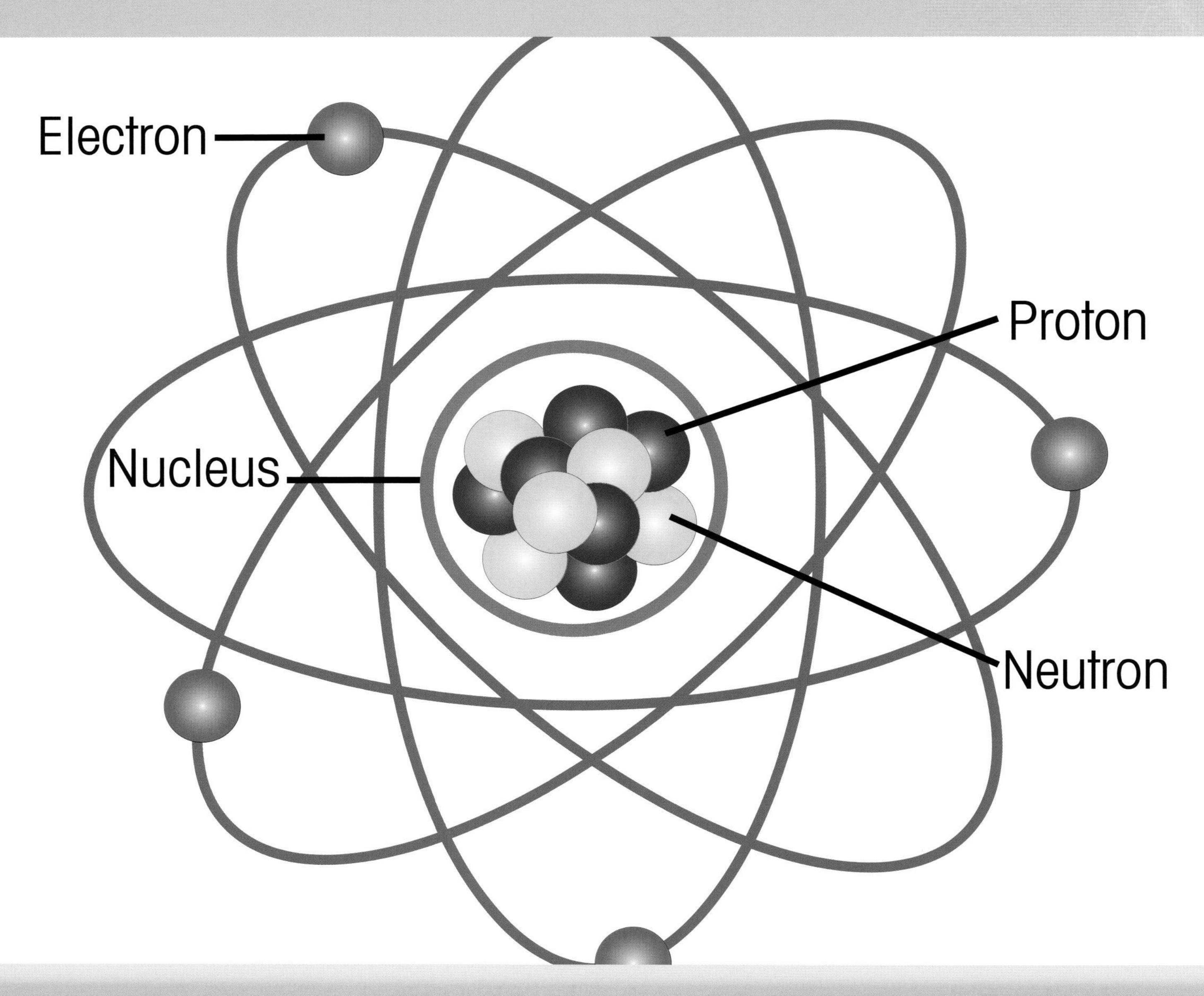

This is a picture of an atom. An atom's nucleus is made of neutrons, shown here in yellow, and protons, here in blue. Electrons, which are colored red in this picture, spin around the nucleus.

Charged Atoms

Neutrons have no charge. Protons have a positive charge, and electrons have a **negative** charge. In most atoms the number of protons is equal to the number of electrons. The negatively charged electrons **cancel** out the positively charged protons. This makes the charge of the atom neutral, or without a charge.

As electrons spin outside of the nucleus, they can sometimes break free from the atoms. When an atom loses an electron, the atom becomes positively charged. This is because the atom now has more protons than electrons. The electron that leaves one atom will then join another atom. An atom that gains an electron becomes negatively charged.

Positive Ion

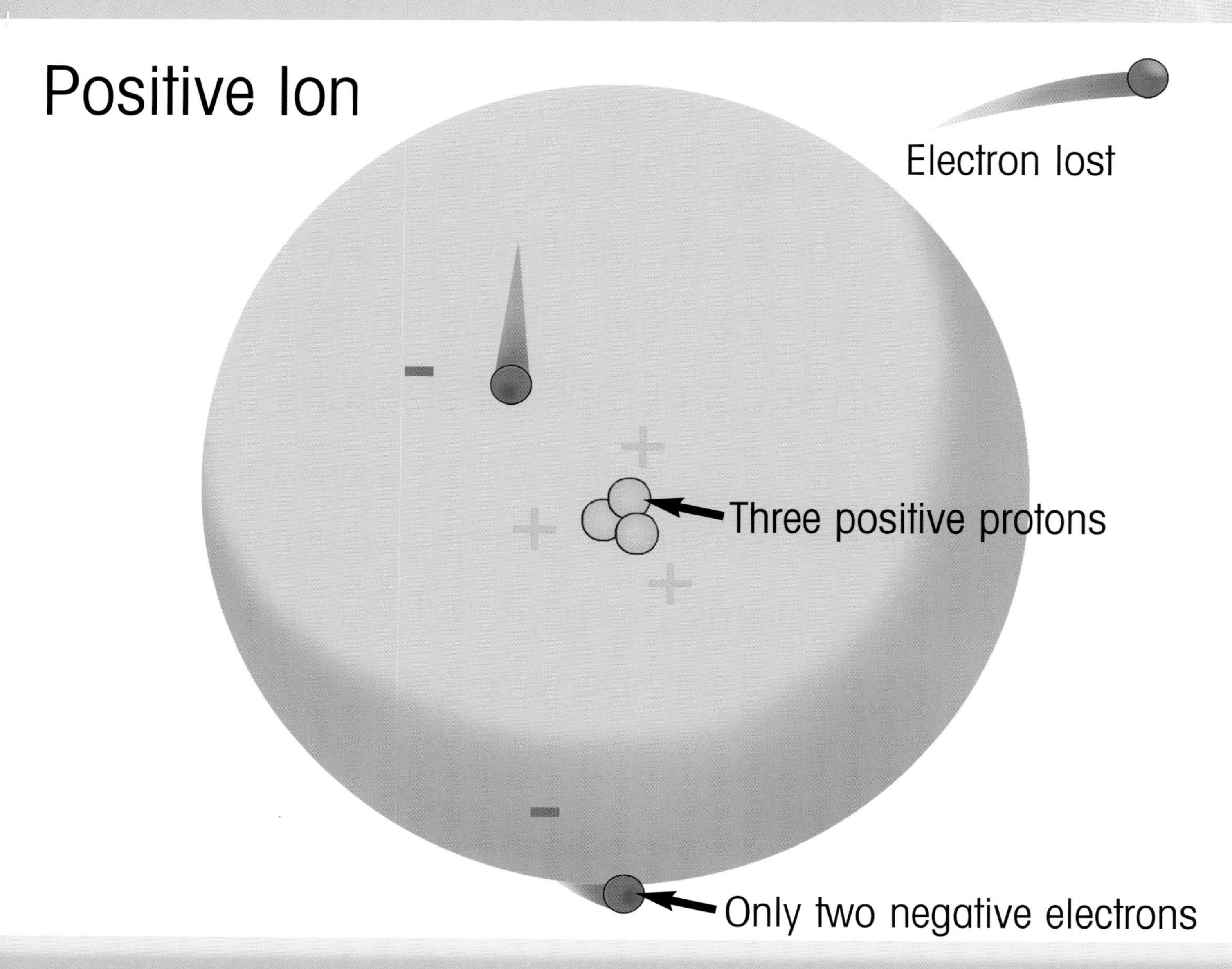

An atom is usually balanced. When it loses an electron, as shown in this picture, it becomes a positively charged ion.

Usually, atoms are balanced and have no charge. When an atom loses an electron, that atom is no longer balanced. The atom will look for another electron to try to balance itself. This is so that the atom will again have an equal number of electrons and protons. The electron also looks for an atom that is missing an electron. Positively charged atoms and negatively charged atoms are attracted to each other. They act like magnets that are attracted to metal. The attraction of positive and negative charges is called electricity. There are two types of electricity. They are **current electricity** and **static electricity**.

Opposite: This boy is using a magnet to attract a piece of metal. *Above:* Magnets attract many metals and can be used as powerful tools. The way magnets pull metal toward them is the same way that ions that have opposite charges pull toward each other.

Current Electricity

A current is something that moves. When an atom loses an electron, sometimes the electron moves directly to another atom. When this happens over and over, it is called current electricity. Current electricity is electricity that moves or travels from one place to another. The distance between the points is called a **circuit**.

Matter that allows current electricity to flow is called a **conductor**. Some **materials** are better conductors than others. Plastic and rubber are poor conductors. Their atoms do not easily lose electrons. Materials that easily lose electrons allow current to flow from one point to another. Many metals are excellent conductors.

Opposite: Electricity is often carried to homes and businesses through wires, such as the ones shown here. *Above:* This is a place where electricity is created, called a power plant. Electricity travels along metal and wire and is stored so it is ready when people need it.

Static Electricity

Static electricity happens when an atom loses or gains electrons and does not balance out right away. In static electricity the electrons do not move as much as the electrons in current electricity. You may have already heard of static electricity. If you blow up a balloon and hold it near a friend's hair, you may notice that the balloon attracts some of your friend's hair. Rub the balloon on your friend's hair. Now hold the balloon near your friend's hair again. Does your friend's hair stand up straight? Some electrons from your friend's hair moved to the balloon. The extra electrons in the balloon are now attracted to your friend's positively charged hair.

Static electricity causes this girl's hair to stand up as she holds the balloon above her head. Try this experiment at home to see static electricity in action.

Electromagnetism

Have you noticed that magnets have the power to stick to objects or attract other objects to them? You may have noticed that the electricity you created in the balloon you rubbed against your friend's head acted very much like a magnet. The reason magnets and electricity act like one another is that they are very closely **related**. We call the relationship between electricity and magnetism electromagnetism. An electric flow through a magnet has an effect on the magnet. If you flow electricity through a magnet, the magnet's powers of attraction will change. Magnets also affect electrical flow. In fact we can use magnets to create electricity.

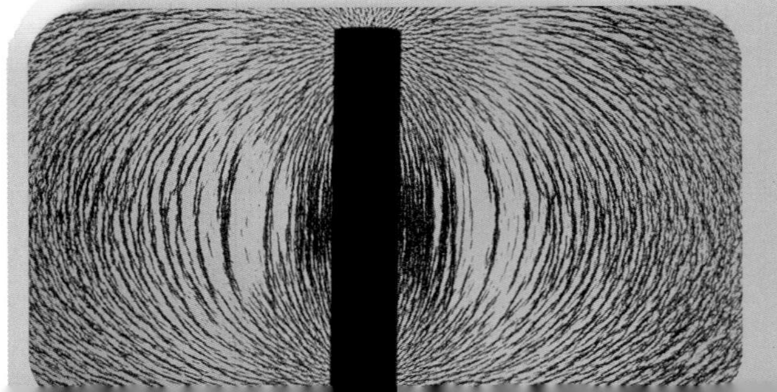

Opposite: Small pieces of metal form a pattern around the two ends of a magnet. *Above:* In power generators, such as this one, water power is used to turn a turbine. The turbine is connected to electromagnets. The electromagnets change the energy from the turbine's motion into electricity.

Making Electricity

People create the electricity that we use. Most of the electricity we use is from **nonrenewable** sources, such as coal and oil. The supply of these **resources** could run out some day. To make electricity from such a source, we often burn the resource to create heat and steam. The steam spins a **turbine**. The spinning turbine sends energy through a magnet, which creates electricity.

We can also make electricity by using wind or water to turn a turbine. Wind and water are **renewable** resources. This means we have a nearly unlimited supply of these resources. Can you imagine life without electricity? The next time you turn on the lights or wear clean clothes, think about how electricity made that possible.

Opposite: Windmills capture the power of wind to make electricity. The wind spins the blades on the windmill. This action in turn powers a turbine. *Above:* Hoover Dam, shown here, creates enough electricity each year to serve more than 1 million people. The Hoover Dam creates electricity using water power.

Experiments with Electricity: Make a Windmill

Supplies Needed:
An 8 x 8 inch (20 x 20 cm) piece of heavy construction paper or posterboard, a ruler, a pencil, a pair of scissors, a thumbtack, a ¼-inch (6-mm) wooden dowel, about 2 feet (61 cm) long

We use many of Earth's resources to make electricity. Energy from the Sun and the wind can be caught and turned into electricity. One way to catch wind energy is using a windmill. Make your own windmill to see wind energy in action.

Step 1 Use a ruler to draw an *X* on the piece of paper. Make your *X* by laying the ruler down so that it passes through opposite corners on the piece of paper. Use the pencil to draw a line. Do the same for the other two corners.

Step 2 At the center of the *X*, draw a 1 x 1 inch (2.5 x 2.5 cm) square. This means that each side of the square is 1 inch (2.5 cm) long. The center of the *X* also should be the center of the square.

Step 3 Starting at one corner, cut along the pencil line until you reach the square. Do the same for the other three corners. Be sure to stop cutting when you reach the square so the piece of paper stays together.

Step 4 Bend the left-hand corner of each triangle into the center, but do not make a fold in the paper. Use a thumbtack to pin the corners to the center and to stick the windmill onto the wooden dowel. Take your windmill outside on a windy day. The energy from the wind makes the windmill blades spin!

Experiments with Electricity: Make an Electromagnet

SUPPLIES NEEDED:
Wire, a yardstick, a long iron nail, a 6-volt battery, scissors, metallic and nonmetallic objects

As you have learned, electricity can be used to make electromagnets. Electromagnets are found in many electrical tools and appliances, including telephones and washing machines. You can make your own electromagnet by following the steps below.

Step 1 Use a yardstick to measure about 2 feet (61 cm) of plastic-coated wire. Have an adult cut away about 1/2 inch (1.3 cm) of plastic at each end to let the wire stick out. Tightly wind the wire many times around a long iron nail.

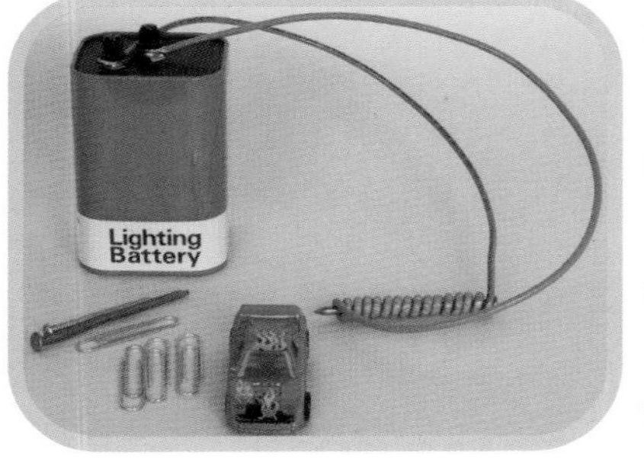

Step 2 Connect the ends of the wire to a battery. See how many objects you can pick up by using your electromagnet.

Glossary

cancel (KAN-sel) To match in force and effect and bring to nothing.

charge (CHARJ) A force like the strength of a magnet.

chemical (KEH-mih-kul) Having to do with matter that can be mixed with other matter to cause changes.

circuit (SER-ket) The complete path of an electric current.

conductor (kun-DUK-ter) Matter that allows current electricity to flow.

current electricity (KUR-ent ih-lek-TRIH-suh-tee) Electricity that moves or travels from one place to another.

electricity (ih-lek-TRIH-suh-tee) Energy that produces light, heat, or motion.

electrons (ih-LEK-tronz) Parts inside an atom that spin around the nucleus. They have a negative charge.

materials (muh-TEER-ee-ulz) What things are made of.

mechanical (mih-KA-nih-kul) Run by a machine or a tool.

negative (NEH-guh-tiv) The opposite of positively.

neutrons (NOO-tronz) Parts with a neutral electric charge found in the nucleus of an atom.

nonrenewable (non-ree-NOO-uh-buhl) Not able to be replaced once it is used.

protons (PROH-tonz) Particles with a positive electric charge found in the nucleus of an atom.

related (rih-LAYT-ed) Like something else.

renewable (ree-NOO-uh-buhl) Able to be replaced once it is used up.

resources (REE-sors-ez) Supplies or sources of energy or useful items.

static electricity (STA-tik ih-lek-TRIH-suh-tee) Electricity in which the electrons creating the electricity move very little.

thermal (THER-mul) Having to do with heat.

turbine (TER-byn) A motor that turns by a flow of water or wind.

Index

Web Sites

Due to the changing nature of Internet links, PowerKids Press has developed an online list of Web sites related to the subject of this book. This site is updated regularly. Please use this link to access the list: www.powerkidslinks.com/eia/electric/